CREATURE FEATURES: CLASSIFY ANIMALS!

IT'S AN INSECT!

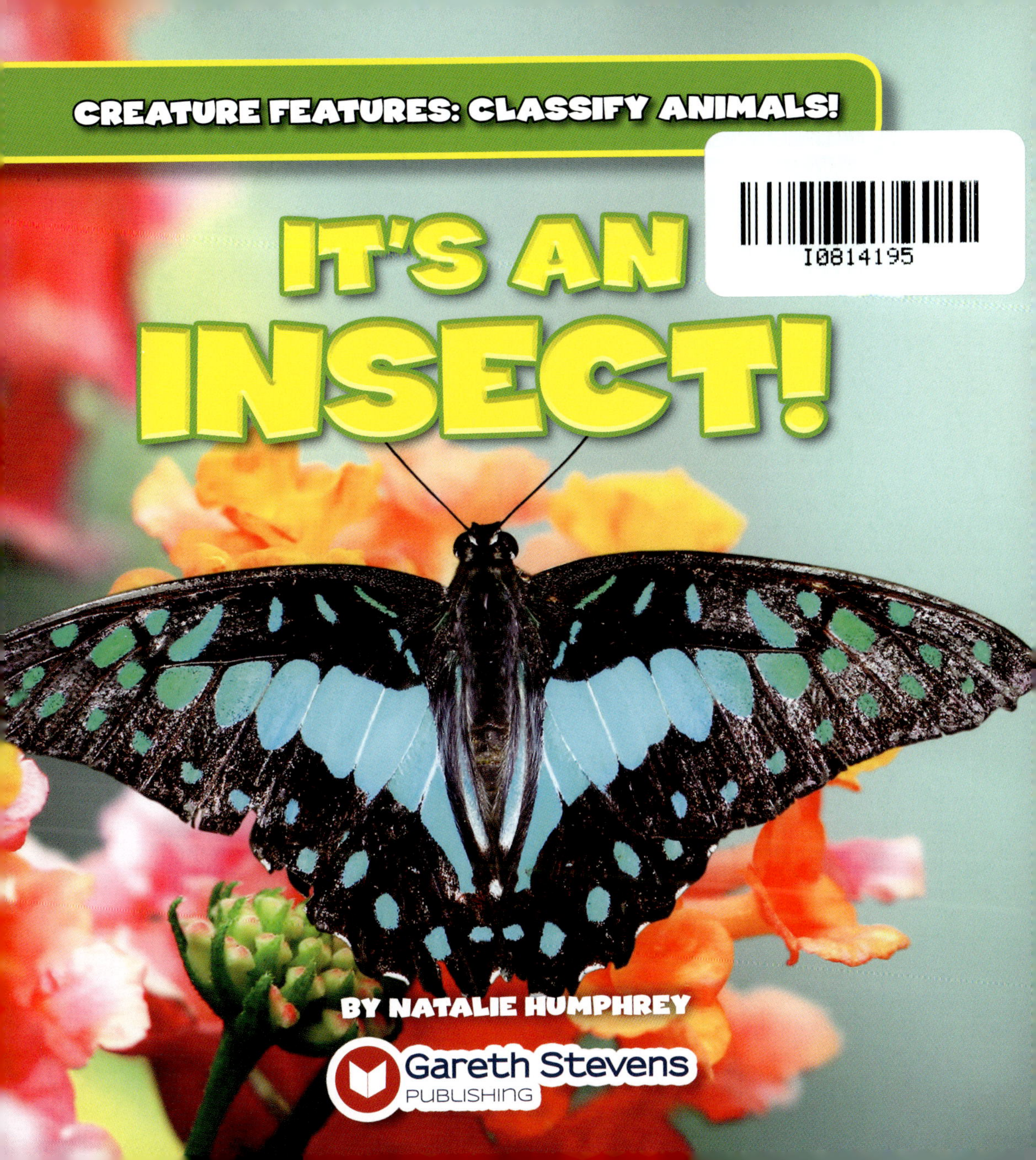

BY NATALIE HUMPHREY

Please visit our website, www.garethstevens.com. For a free color catalog of all our high-quality books, call toll free 1-800-542-2595 or fax 1-877-542-2596.

Library of Congress Cataloging-in-Publication Data
Names: Humphrey, Natalie, author.
Title: It's an insect! / Natalie Humphrey.
Description: Buffalo, New York : Gareth Stevens Publishing, [2025] | Series: Creature features: classify animals! | Includes index.
Identifiers: LCCN 2024008537 (print) | LCCN 2024008538 (ebook) | ISBN 9781482466966 (library binding) | ISBN 9781482466959 (paperback) | ISBN 9781482466973 (ebook)
Subjects: LCSH: Insects–Juvenile literature.
Classification: LCC QL467.2 .H848 2025 (print) | LCC QL467.2 (ebook) | DDC 595.7–dc23/eng/20240229
LC record available at https://lccn.loc.gov/2024008537
LC ebook record available at https://lccn.loc.gov/2024008538

Published in 2025 by
Gareth Stevens Publishing
2544 Clinton Street
Buffalo, NY 14224

Designer: Andrea Davison-Bartolotta
Editor: Natalie Humphrey

Photo credits: Cover, p. 1 Kurit afshen/Shutterstock.com; p. 5 irin-k/Shutterstock.com; p. 7 Ikordela/Shutterstock.com; p. 9 KRIACHKO OLEKSII/Shutterstock.com; p. 11 (bottom) Hawk777/Shutterstock.com; p. 11 (top left) asharkyu/Shutterstock.com; p. 11 (top right) Kirschner/Shutterstock.com; pp. 13 (bottom), 15 Tomasz Klejdysz/Shutterstock.com; p. 13 (top) aaltair/Shutterstock.com; p. 17 Stephen Farhall/Shutterstock.com; p. 19 Darkdiamond67/Shutterstock.com; p. 21 (bottom) ChaiyaTN/Shutterstock.com; p. 21 (top left) mchin/Shutterstock.com; p. 21 (top right) ArtLovePhoto/Shutterstock.com.

Printed in the United States of America

CPSIA compliance information: Batch #CS25GS: For further information contact Gareth Stevens, New York, New York at 1-800-542-2595.

CONTENTS

Boldface words appear in the glossary.

Is That an Insect?

When people think of animals, they don't often think of insects. But insects are the most common animal found on Earth! Scientists believe there are more than 900,000 species, or kinds, of insects alive today.

What Do They Look Like?

An insect's body has three main segments, or parts. The first part is the head. Insects usually have antennae, or feelers, on their heads. The second segment is the **thorax**. Connected to the thorax are the insect's six legs. The final segment is an insect's **abdomen**.

THORAX
ABDOMEN
HEAD
ANTENNAE

The Exoskeleton

Insects have a hard, shell-like outer covering called an exoskeleton. An insect's exoskeleton does the same job that your bones do. The exoskeleton gives an insect's body **support**. It helps to keep an insect's **organs** safe. The exoskeleton also helps an insect move!

Insect Wings

Many insects have wings. Most have two pairs. Insect wings are often thin and **transparent**. Some insects, like butterflies and moths, can have colorful wings. Their wings come in many different shapes and **patterns**.

Insect Eggs

Most insects lay eggs. Some insects lay a few eggs at a time. Others lay hundreds or thousands! Some insects stick their eggs to the leaves and stems of plants. Other insects, such as mosquitoes, lay their eggs in water.

Live Babies

Not all insects lay eggs. Some mother insects keep their eggs inside of their bodies. The eggs **hatch** inside of the mother and the babies are born live. Other insects, like some aphids, don't hatch from eggs and are born live.

APHIDS

Larvae

Most insects hatch into larvae that look much different than their adult form. A larva may not have legs and look more like a worm. Many larvae eat plants, but some eat other insects. These larvae will even eat other larvae!

Metamorphosis

Most insects go through **metamorphosis** before becoming adults. Some kinds, such as crickets, **shed** their skin as they grow into adults. Others build a **cocoon** around their body. The larva's body breaks down in the cocoon, and it grows legs and often wings!

METAMORPHOSIS

Insect Food

Many adult insects eat mostly plants. Some insects, including ants, eat nearly anything from fruit to the eggs of other insects. Female mosquitos and bedbugs only eat the blood of other animals. Insects share a lot of features, but can also be very different!

GLOSSARY

abdomen: The final part of an insect's body that includes most of its organs.

cocoon: A case an insect larva makes around itself in which it undergoes metamorphosis.

hatch: To break open or come out of.

metamorphosis: The body changes an insect goes through from larva to pupa or nymph to adult.

organ: A part inside an animal's body.

pattern: The way colors or shapes happen over and over again.

shed: To lose skin.

support: To hold up.

thorax: The middle part of an insect's body where the wings and legs are connected.

transparent: Letting light shine through.

FOR MORE INFORMATION

BOOKS

Mound, Laurence. *Insect.* New York, NY: DK Publishing, 2023.

Thomas, Isabel. *One Million Insects.* London, UK: Welbeck Editions, 2021.

WEBSITES

Britannica Kids: Insect
kids.britannica.com/kids/article/insect/353292
Learn more facts about insects and what makes them different from other animals.

National Geographic Kids: 25 Cool Things About Bugs!
www.natgeokids.com/uk/discover/animals/insects/15-facts-about-bugs/
Discover more facts about insects.

INDEX